OPERATION SUSTAINABLE HUMAN

A FOUR-STEP SCIENTIFIC GUIDE TO COMBAT CLIMATE CHANGE

(high impact made simple)

BRIGHT IDEAS TOWARDS A HEALTHY
AND SUSTAINABLE FUTURE

1st Edition

Illuminate Press, Vancouver

Produced by
Chris Macdonald, Paul Macdonald, James Tocher, and Roger Williams.

ISBNs

Paperback	978-1-7752528-3-2
Electronic	978-1-7752528-4-9
Audio	978-1-7752528-5-6

This book is available as a paperback, eBook, and audiobook.
To find out more please contact the author.

THE AUTHOR

Chris Macdonald is a scientist and author. He wrote this book as part of a nonprofit sustainability project. To get in touch and to see Chris's latest projects, you can visit his Instagram page: *@ChrisMacdonaldOfficial*

For every paperback sold, ten trees will be planted. To achieve this we have become an official partner with *Trees.org*. Any additional profits will also be used to help combat the climate and ecological crisis.

SPECIAL THANKS

In alphabetical order:

Adelina Suvagau, Alan Burgin, Austin Andrews, Ben Rew, Ben Woodhouse, Brad Goodman, Claire Macdonald, Dad, Daryl Hatton, David Rew, David Shortt, Francis Lecompte, Glen Samuel, Grace Rew, Gráinne Rew, Greg Durrell, Hossein Martin Fazeli, James Tocher, Jamie Macdonald, Jeff Cullen, Jenny Macdonald, Karen Pilkington, Katherine Maunder, Kevin Gurney, Kevin Jacques, Lauren Fry, Lucy Ogilvie-Grant, Marion Misériaux, Mike Coto, Mum, Nicole Gallienne, Nicolas Provencal, Paul Macdonald, Pippa Burgin, Pi Ware, Roger Williams, Roisin Clarke, Tom Wylde, Trevor Meier, Whitney Radforth, William Bartlett.

And, a huge thank you to those who demand climate action from their political representatives.

CONTENTS

CHAPTER 1

MISSION

Over a decade has passed since I launched my first climate change project; my goal at the time was to raise awareness. Since then, great strides have been made. Yet, while I feel that more people are becoming aware of the severity of the problem, what we should do about it, as individuals, is far less understood.

If you search online for possible climate action items you can find some rather hefty to-do lists (such as *101 things you can do to save the planet*). When studying these lists, I noticed they tended to be inefficient, and they often missed out some of the most significant offenders. In some cases, addressing a single item not on the list would make a greater impact than if you completed the entire list of recommendations.

I felt that it should be made a lot easier and simpler for people, and that we could be far more impactful with an optimized strategy. I didn't like the thought that well-meaning people might be carrying out many time-consuming tasks that—in all honesty—make near-zero difference. I also didn't like the thought that the biggest pollutants are not only going ahead somewhat unchecked, they are rapidly expanding.

This book seeks to remedy these concerns. It is a straight-talking, clearly referenced, scientific guide. It will not pretend that the climate crisis is not all that bad nor will it propose that simply remembering to turn off the lightbulb is sufficient climate action. It will tell it like it is. And it will only offer action items that make a highly significant difference. No speculative theories. No time-wasting. No BS.

We will not dwell on how things could have been managed better. Instead, we will assess the situation that we find ourselves in and explore how best to proceed in line with the united science.

I appreciate that it can often be a confusing and overwhelming subject matter, and so rather than bombarding you with 100+ things to do, you'll be offered a deeply considered, optimized list of four steps, that tackle the root causes and account for the vast majority of the problem.

As a friendly piece of advice, I'd like to recommend that the reader refrain from feelings of shame and regret. You will likely discover ways in which you are negatively contributing to global warming, and at a rate that you hadn't anticipated. The intention is not to make you feel bad. The genuine intention is to help as much as possible. So please read on with the confidence that you will also find sufficient, high impact, actionable solutions—which are actually rather straightforward.

While we may have previously made mistakes, the next chapters of our lives have not yet been written. And we can decide which role we'll play. There is still time to be on the right side of history.

In short, the mission of this book is ambitious yet simple: to make it as easy as possible to make a highly significant positive difference in a short timeframe—to assist those who want to do their part to help combat the climate and ecological crisis.

I hope it helps.

Thank you.

CHAPTER 2

BRIEFING

Before exploring the solutions, there are some very important points to clarify when it comes to global warming. Let's begin with an incredibly brief summary of how it occurs.

The main cause of global warming is the greenhouse effect. This is where some of the gases in our atmosphere—such as carbon dioxide, nitrous oxide, and methane—let sunlight in but prevent some of the heat from escaping. As a result, the average surface temperature of our planet will continue to rise as more of these gases accumulate in the atmosphere.

There are many areas that we could explore when it comes to global warming and harmful pollutants, but to keep the book as concise as possible, and solution-focused, I have decided to stick to the three most pertinent facts.

FACT 1:
HUMANS ARE CAUSING GLOBAL WARMING

It is crucial to bust the incredibly dangerous myth that global warming has nothing to do with human activity.

For starters, the leaders of every country in the world signed the Paris Agreement: an agreement where nations work together to mitigate the human causes of global warming [1] [2] [3]. Secondly, when reviewing the data from over 4,000 climate research papers, it was concluded that there is now a near 100% agreement among scientists that human behavior is responsible for global warming [4] [5]. NASA, the National Academy of Sciences, and every major scientific organization, recognizes this [6].

This is understandably a tough pill to swallow. It isn't nice to know that we are the problem, which in part, can explain the scale of denial and the spreading of false information. An example of which, is the false claim that human emissions are insignificant compared to volcanic eruptions.

When accounting for the world's volcanoes—both above and below the sea—we get an annual total of about 200 million tons of CO_2, which is indeed a significant amount. However, the emissions from the world's volcanoes are less than a single percent of the emissions produced by modern-day human activity [7] [8].

FACT 2:
POLLUTION SERIOUSLY HARMS AND KILLS

There are many ways that pollution can impact our health. And some of them are starting to become more well known—such as the increased risk of respiratory diseases as a result of air pollution [9] [10] [11]. However, there are many more health impacts that we are only just beginning to understand.

Plastic, for example, is also a very harmful pollutant, and it is increasingly difficult to avoid. When analyzing the world's most popular brands of bottled water, researchers discovered pieces of plastic in more than 90% of them [12]. In a single bottle of Nestlé Pure Life, concentrations were as high as 10,000 plastic pieces per litre of water [13]. And, as plastic is also found in animals [14] [15] [16] and processed food [17] [18] [19], it's no surprise that it is also found within us [20] [21] [22] [23]. For example, BPA—a chemical commonly used in consumer plastics—is now found in more than 90% of the U.S. population [24]. This is hugely concerning as BPA can mimic estrogen, and as a result, it can significantly lower a male's sperm count and cause erectile dysfunction, among other sexual health problems [25] [26]; BPA is also linked to heart disease, diabetes, and cancer [27] [28] [29].

Beyond plastic, the various other harmful modern-day pollutants, and the resulting changes in the climate, already account for millions of deaths each year [30] [31]. And the latest studies warn that if we don't change our ways the death toll will skyrocket [32] [33], not to mention the many millions who will be forced to become climate refugees [34] [35].

The devastation comes from multiple compounding crises: coastal flooding, ocean acidification, forest fires, power outages, food shortages, fresh water depletion, infections, disease, malnutrition, the loss of livelihoods, and the loss of homes [36 37 38 39].

Put simply, pollution is not a victimless crime. It kills. And it kills on a horrific scale.

FACT 3:
RAPID DRASTIC CHANGE IS REQUIRED

In the succinct and sagacious words of Greta Thunberg, "Our house is on fire." [40] This may seem like a rather poetic exaggeration but we are indeed in a perilous position and time is against us.

Climate change is not an abstract theoretical threat as we are currently experiencing the 6th mass extinction and we are rapidly destroying the vital ingredients of life [41 42 43].

Half of the Earth's original forest cover has already gone, and every year we destroy millions of acres more [44 45 46]. We have already annihilated over 80% of wild mammals and every year we continue to render thousands of species extinct [47 48 49 50]. In the past few decades alone, we have lost half of the world's coral reefs [51] [52] [53]. With our current rate of soil degradation we only have 60 years left of arable farming [54] [55]. Our current fishing practices may no longer be possible in 30 years [56] [57]. And the Arctic region could have its first ice-free summer in just 10 years [58] [59]. Therefore, the effects of overconsumption, pollution, and the resulting changes in our climate, are already deadly and they will rapidly become a lot worse.

It is much nicer to think that the climate and ecological crisis hasn't yet begun and that we still have plenty of room left to play with. In reality, however, we have been on borrowed time for decades. We went hurtling beyond the safe atmospheric carbon threshold in the late '80s and we have only been accelerating since then [60] [61]. Therefore, we are already in a deadly deficit and we are quickly diving deeper into debt.

Just to clarify, this means that not only do we need to immediately stop adding to the problem, we also need to rebuild ecosystems to capture the excess emissions.

It is also important to note that we are on the verge of several tipping points. This means that once we pass a certain point there will be a chain of amplifying feedback loops that can accelerate climate change well beyond our control. An example of this is the massive greenhouse gas release that comes from thawing Arctic permafrost—which is currently thawing at an alarming rate, 70 years earlier than predicted [62] [63] [64] [65]. Also, as the snow and ice melt, it exposes darker land surfaces which absorb more sunlight, and therefore further accelerate global warming [66].

You might be thinking that if the climate crisis is this bad, why aren't thousands of scientists warning us? They are, and they have been for decades [67] [68] [69]. Most recently, more than 15,000 scientists from all around the world issued a global warning that we need rapid, extreme change in order to save life on Earth [70]. And in the latest report from the United Nations, it warned that we only have around 10 years to cut emissions by 50% to avoid utter catastrophe [71].

And yet, consumption is increasing and emissions are rising.

The heart-wrenching truth is that those suffering the most are often the least responsible [72]. And perhaps the biggest victims of business-as-usual are yet to be born.

In summary, it is not a far off, distant crisis; it is here

now, in this very moment. It is already irreversible, in so far as millions of people and countless opportunities have already died. What's on the horizon, if we don't change course, is exponentially worse. We are in the midst of a deadly, yet optional, crisis. We can end it with rapid and drastic change. In other words, what we decide to do or don't do, matters. A lot.

When examining what contributes to the climate and ecological crisis, it is apparent that there are a few offenders that account for the vast majority of emissions. This means that by prioritizing key areas, we can make massive improvements in a very short amount of time. Please read through the following four steps to see how you can be part of a genuine solution. Each chapter is concise, easy-to-read, and ends with a single, highly impactful action item, backed by science.

Yes, our current situation is bad. Really, really, bad. Far worse than most people are aware. But, what stops it becoming astronomically, unimaginably worse, is the fact that we have proven solutions available to us. Here are some of the most potent ones.

CHAPTER 3

FOOD

The biggest contributors to the climate and ecological crisis highlight just how disconnected we have become from the sourcing and manufacturing of our goods. A prime example of this is our food; there can be hundreds of ingredients in a single piece of processed food, and you'd do very well to recognize them in a lineup, let alone say where they came from.

Animal agriculture may be the epitome of consumer disconnect; a lot has happened before that plastic-wrapped section of meat was placed on the supermarket shelf. And those prior processes are devastating for our environment.

First of all, we have the enormously inefficient use of

land. Not only to raise the livestock but to grow feed for the livestock. Animal agriculture is the largest use of land globally [73, 74], and it is responsible for over 90% of Amazon destruction [75]. An acre of rainforest is cleared every second [76, 77], and animal agriculture is the number one culprit [78, 79, 80].

In fact, livestock and livestock facilities now cover over one-third of the planet's land surface [81, 82]. To help put that into perspective, all of the paved surfaces, roads, and buildings combined cover less than a single percent of the planet's land surface [83].

Not too surprising then that we could feed well over 10 times as many people with the same amount of land if we stopped consuming animal products [84, 85].

Beyond the incredibly inefficient use of land and the gigantic loss of carbon-capturing trees, animal agriculture also uses monumental amounts of water. A single cow can drink 100 gallons every day [86], and we also have to factor in the water used to cultivate their feed crops, and the water used to frequently clean the animal's facilities. All-in-all, it takes around 2,500 gallons of water to produce a single pound of beef [87, 88]. You might be currently taking shorter showers to help conserve water, yet just two pounds of beef is the equivalent water use of showering every day for an entire year.

Since cows consume so much, they produce an enormous amount of waste. A single farm with 3,000 dairy cows produces more waste than a city of 400,000 people [89]. In the U.S. alone, livestock produces over 100,000 pounds of waste, each and every second [90].

It is also wise to consider the greenhouse gas emissions. Animal agriculture is one of the world's leading carbon polluters, to the extent that simply avoiding beef will reduce your carbon footprint more than if you stopped driving a car [91]. Beef and dairy farms are also one of the primary methane polluters, as a cow can produce hundreds of litres of methane in a single day [92] [93]. And methane is over 20 times more potent than CO_2 [94] [95]. Cows are also responsible for two-thirds of the world's nitrous oxide pollution [96], which is devastating when we consider that nitrous oxide has nearly 300 times the global warming effect of CO_2, and it stays in our atmosphere for over 100 years [97].

The impact of animal agriculture is so profound that if you simply reduce your intake of, or found a replacement for, beef alone, you'd make a highly significant improvement to our current climate. In fact, it was recently calculated that if Americans switched from eating beef to beans, the U.S. could still come close to meeting its greenhouse emission goals, despite withdrawing from the Paris Agreement [98 99].

As noted by environmental researcher Helen Harwatt, "The real beauty of this kind of thing is that climate impact doesn't have to be policy-driven." [100] This is an empowering time as we can personally make a highly significant difference, providing we listen to the united science and are willing to step out of our comfort zones and embrace the available solutions.

At this point, it is worth clarifying a couple of points, starting with the myth that you can't get enough protein if you avoid animal products. This is incredibly inaccurate. Not only are their many raw sources of plant protein (lentils, quinoa, spirulina, hemp, peas, beans, seeds, nuts, and so on), there are now plant-based alternatives that match or beat the protein content for a range of popular items: protein powders, burgers, sausages, mince, milk, etc. This is why we can easily find elite athletes, who follow a plant-based diet, in a range of disciplines: from ultramarathon runners and triathletes to professional bodybuilders and Olympic weightlifters [101] [102] [103] [104].

A second aspect that is important to clarify, is just how much food the animals eat. The entire human population eats around 20 billion pounds of food every day, but the cows alone, eat over ten times more [105]. By cutting out the 'middleman', we'd be able to prevent avoidable emissions, capture existing emissions, and feed a great deal more people [106]. Hence why the Alliance of World Scientists—a group of over 10,000 scientists from over 100 countries—recently urged the public to switch to a plant-based diet [107]. And even better if you can grow your own or support local permaculture projects.

Since we have inherited certain systems, it is important to take a step back and really think about what we are doing. We are artificially inseminating and raising billions of living beings. While keeping them in unnatural confinement, we transport monumental amounts of scarce, valuable resources to them, only to eventually eat them—with an incredibly poor return on investment. It is inefficient, unnecessary, and extremely unsustainable. But the problem is not 'evil consumers' nor a lack of empathy. I believe the problem lies with unprecedented levels of consumer disconnect and a huge lack of awareness.

When we visit the supermarket, there is an illusion of sustainability and unending abundance, but outside of those walls, vital biological systems are being destroyed at ever-increasing rates. While we used to respond to the source within nature, we now respond to products within narratives.

ACTION ITEM No 1: SWITCH TO A PLANT-BASED DIET

* Or at least try to avoid or limit lamb, beef, and dairy (the most significant polluters). Start big or small. Just start somewhere.

* Additional resources, support, and advice for every action item suggested in this book are available on my Instagram page: *@ChrisMacdonaldOfficial*

Switching to a plant-based diet has rapid, measurable,

highly significant results. In a single month, you personally would have saved 600 pounds of CO_2, 900 square feet of forest, 1,200 pounds of grain, and 33,000 gallons of water [108]. Just by modifying your diet you can stop funding the leading cause of deforestation [109] [110], ocean dead-zones [111] [112] [113], habitat destruction [114] [115] [116], species extinction [117], water pollution [118], methane pollution [119], and nitrous oxide pollution [120] [121].

In short, it is the single biggest way to reduce your negative impact on our planet [122] [123] [124] [125] [126] [127] [128] [129].

CHAPTER 4

TRANSPORT

Humans increase carbon dioxide levels with a brutal two-pronged attack. We increasingly destroy natural CO_2 absorbents (such as forests), and we use systems and tools that increasingly add massive amounts of CO_2 into the atmosphere [130]. In other words, we are plugging the drain and running the tap at the same time.

Transport and travel are one of the main sources of CO_2 [131] [132], and for many nations, it is the single biggest source of carbon emissions [133]. But, in addition to the potent global warming effect of our vehicles, we also have the direct health impact from their pollution.

The pollution from our vehicles contaminates our

crops [134], soil [135], water [136], and the air that we breathe. While some of the health risks associated with air pollution are becoming more well-known such as the previously mentioned increased risk of respiratory diseases [137]—it can harm us in many other ways.

Air pollution can adversely affect the nervous system, kidney and liver function, skin condition, eye-sight, blood pressure, reproductive systems, and the immune system [138] [139] [140] [141]. It is incredibly dangerous as the fine particles can lodge deep into your lungs [142] [143], as well as interfere with your blood's ability to transport oxygen [144]. Air pollution is also linked to higher rates of miscarriages, dementia, heart disease, strokes, and cancer [145] [146].

Recent research has also shown that infants are particularly vulnerable to air pollution, to the extent that it can result in lower birth weight, behavioral problems, and even a significantly lower IQ [147] [148] [149] [150].

One might assume that air pollution is only really dangerous in the most polluted cities—Jakarta, Mumbai, Delhi, Beijing, and so on—however, the reality is that over 90% of the world's population now live in areas with unhealthy air quality [151] [152]. And already, it is causing millions of premature deaths [153] [154] [155]. Hence why the World Health Organization state that air pollution is a public health emergency [156].

It is harrowing to contemplate, that each time you start up your vehicle, you are seriously affecting the health of your family and local community. If exhausts were to bellow out thick red smoke and people were to drop down in the streets as soon as it touched them, then we would have abandoned fossil fuel travel a long time ago. But they don't. Instead, the emissions steadily build up and slowly harm us in a number of discrete ways. And now nearly every town and city has been poisoned. As they say, it is the slow knife—the one that quietly takes its time—that cuts the deepest.

And, if we are truly serious about clean travel, we also have to reconsider flying.

Aviation is disastrous for the environment. In fact, if the emissions from aviation were attributed to a single country, that country would immediately become one of the world's top polluters [157].

Even a modest round-trip flight produces a metric

ton of CO_2—per passenger [158]. What's more, burning jet fuel also releases soot, sulphate, and nitrous oxides [159]. And unfortunately, high-altitude emissions have a far greater impact on the climate [160] [161].

The tropospheric ozone and vapour trails produced by air traffic have incredibly potent warming effects in a relatively short amount of time [162]. Which is one of the reasons why scientists state that if we measure the more immediate effects, then planes may account for more warming than all of the cars on the roads, despite there being a huge difference in the number of users [163].

The impact of aviation is a particularly troubling issue as there is an ever-increasing demand for flights. If we don't curb our usage, there will be double the number of planes in the sky in just 15 years [164]. And due to the international nature of flying, it can be kept out of a nation's carbon budget [165] [166], meaning that it is even more critical that change comes from the individual.

To put it simply, we need to keep fossil fuels in the ground, and flying is essentially another fossil fuel industry [167]. Therefore, if we are serious about limiting global warming to 1.5°C above pre-industrial times (as outlined in the Paris Agreement [168]), then we also need to avoid flying.

ACTION ITEM No 2:
AVOID FOSSIL FUEL TRAVEL

Cleaning up the way you travel, where possible, is an incredibly positive contribution to your planet, your community, and yourself. If we can make the switch to sustainable public transport, cycling, and walking, then we cut a massive amount of emissions, protect our children, and make our communities a lot quieter, cleaner, and safer.

CHAPTER 5

SUPPLIES

There are many days in the year that we are sure to keep track of: public holidays, anniversaries, birthdays, and so on. However, one very important day often goes by under the radar.

Overshoot Day is when resource consumption has exceeded the sustainable limit. That is to say that once we arrive at Overshoot Day, we have used up the amount of resources that our planet can regenerate in a year. When we pass this day, we are in ecological debt.

If we were a sustainable species, we would never see Overshoot Day, as we would only be using a fraction of the earth's annual resources. However, we currently see Overshoot Day at the end of July, and each year it arrives earlier [169].

Therefore, for us to simply maintain our current lifestyles, we would need two planet earths.

Before discussing how and why we consume so much, it is important to bust another incredibly dangerous myth: that we are unsustainable because planet earth is overpopulated. This is dangerous because it promotes a highly pessimistic, 'why bother' attitude. And it is particularly dangerous because it is so appealing. It lets people off the hook as it implies that we have gone beyond the point of no return. It is the comfortable perspective as it can justify inaction. As a result, it can make the current situation significantly worse. Thus, it is even more important to remember that it is only a myth.

With over 7 billion people on our planet, there are indeed a lot of mouths to feed. But even with our incredibly inefficient systems, we produce enough food to feed 10 billion people [170]. What's more, we are also choosing to raise and feed over 70 billion farm animals every year [171]. While overpopulation is something we should certainly keep track of, it is not the problem.

The current problem is over-consumption. Despite the modest needs of us Homo sapiens, we are unnecessarily consuming well beyond what we can maintain.

A great example of unsustainable consumption is the fashion industry; with paid influencers, social media posting, rapid trend rotations, and powerful marketing ploys, we now purchase new clothing at unprecedented rates. As a result, 100 billion new garments are created every year [172 173], and manufacturing is ramping up [174].

We are buying more items of clothing, wearing them less often, and in the vast majority of cases, not recycling them. Worldwide clothes are now being discarded at the rate of an entire landfill truck every single second [175]. Accordingly, fashion is one of the fastest growing categories of waste [176].

In addition to this, a great deal more waste is kept off the record. For example, it was recently discovered that Burberry burned almost 40 million dollars worth of clothing. Rather than selling it at discounted rates, the brand wanted to destroy the unsold items to maintain exclusivity [177]. And there is growing suspicion that this is far from an isolated incident. Fashion activist, Orsola de

Castro, claims that burning has been fashion's dirty secret for decades [178].

Although we purchase so much clothing, we rarely consider the environmental impact. It takes over 2,000 litres of water to produce enough cotton to make a single t-shirt [179]. And the cotton industry uses more pesticides than any other crop [180]. The clothing industry is also a major source of greenhouse gases [181], it is a significant contributor to deforestation [182], and it is the second largest polluter of clean water [183].

Since we are talking about unsustainable consumption, we really ought to mention plastic again.

Plastic purchasing has also reached unprecedented levels; in the last ten years alone, we have produced more plastic than the previous 100 years combined [184]. And we only recycle 5% of it [185]. We have now reached a level where enough plastic is thrown away each year to circle the Earth four times [186]. And it can take 1,000 years to degrade [187].

As a result, plastic is everywhere. When previously discussing the health implications of pollution, we saw that plastic can find its way into animals [188], food [189], water [190], and us [191]. But even that doesn't do justice to how much plastic has invaded our planet.

Explorer Victor Vescovo recently broke the record for the world's deepest dive. And when he descended over 10 kilometres into the extreme depths of the uncharted ocean, what did he find? Plastic [192].

We even discover plastic before we are released into the world, as over 200 chemicals, including BPA, have been found in umbilical cord blood. This means that plastic chemicals are now present during incredibly vulnerable stages of fetal development [193] [194].

There is now so much plastic thrown away, that we have gigantic plastic wastelands floating in our oceans. Plastic Island is the name given to one of the largest masses. And while one might argue that the label of island may be a touch overdramatic, Plastic Island is now three times the size of France [195].

And despite this, corporations are ignoring the problems and continuing full steam ahead, with the fossil fuel industry aiming to increase plastic production by 40% within the next decade [196].

With so many items being purchased and discarded, it

is also important to consider how they get from A to B.

The chances are that the majority of the items that fill your home were sourced or manufactured in another country, which results in a massive amount of emissions via international shipping [197]. One of the key issues with this is that, as with flying, the emissions take place between countries. Thus they are not considered as national emissions and are therefore not part of the national reduction targets listed in the Paris Agreement [198]. Which is a huge issue as the carbon emissions from international shipping are equivalent to the emissions of an entire country [199].

It also appears that, as we consume more items, our standards loosen. We have seemingly adopted a transient mindset in which we almost expect certain items not to last. And we no longer expect them to be built in a way that facilitates repair work. Accordingly, t-shirts, jeans, and even smartphones are usually discarded within a few years [200].

And, as we consume and discard more often, we start to ask fewer questions. We are no longer asking where or how our items are made. Which, aside from a massive decrease in skills, resilience, and self-sufficiency, it also further loosens standards, and can further perpetuate consumption.

Over-consumption is a dense area, with many compounding negative aspects, and so rather than suggesting a comprehensive list of micro-consumerist dos and don'ts, I'll simply advise a more general rule of thumb: consume less. A lot less. And when you do consume, please try to investigate the complete cost, as consumption impacts far more than our pockets.

ACTION ITEM No 3: CURB YOUR CONSUMPTION

* In particular, avoid high-emission and limited-use products.

We don't have to obediently follow unsustainable lifestyles and consumer trends simply because they have become the status quo; we can instead opt to make highly considered, compassionate, conscious decisions. We can break through the conditioning—we can dare to be different.

Cutting down on consumption can save time and money. It significantly reduces our environmental impact, and it can be outstanding for our mental and physical health. Focusing less on superficial, material consumption will likely lead us to become more grateful with what we do have, and perhaps we'd begin to focus more on what is really of value.

In short, limited-use consumables are an incredibly foolish use of our precious, finite resources; it is wise to live well within planetary boundaries. Unsustainable practices—by definition—must come to an end. It is time to rapidly curb consumption and dispose of the disposable era.

CHAPTER 6

SYSTEMS

In addition to the greenhouse effect, we also have to consider the whitehouse effect. And by that I mean we have systems in place that are not only grossly inadequate at responding to climate change, they also facilitate it and, I'd argue, they inherently promote it.

While the greenhouse effect relates to the harmful consequences of certain gases in the atmosphere, the whitehouse effect relates to the harmful consequences of money in politics.

In particular, there are two ways that money in politics impacts our climate: funding and focus. Let's begin with funding.

It is wise to acknowledge that those with vast sums of money can make enormous donations to campaigns which can leave us with heavily compromised politicians. To put it bluntly, many politicians are accepting bribes.

While this might appear as though we are departing from scientific rigour and entering into the realm of conspiracy theory, we now have overwhelming evidence that corporate investment can have some devastating consequences for climate action.

The top fossil fuel firms spend hundreds of millions of dollars on advertisements claiming that they support climate action [201], while at the same time, they spend hundreds of millions lobbying to block climate action policies [202]. This is a textbook example of the infamous doublespeak: misleading communications that distract and deceive.

Another example is that while so-called pro-climate, fossil fuel corporations spend a tiny fraction of their expenditure on low-carbon technologies, over $100 billion will be spent on more fossil fuels [203].

Corporate funding of this nature can help to explain some highly illogical and counter-intuitive political decisions. For example, given the scientific consensus [204] [205], and the fact that every country signed the Paris Agreement, it seemed bewildering that the U.S President withdrew from the agreement in 2017 [206]. And the decision was accompanied by greater confusion when the President added that the U.S. "will continue to be the cleanest and most environmentally friendly country on Earth."[207]–a highly perplexing statement coming from the biggest carbon-polluting country in history [208].

However, when we consider corporate funding, it begins to make sense. As not only is the President heavily funded by fossil fuel corporations [209], so too are the 22 senators who urged Trump to withdraw from the agreement [210]. Senator Ted Cruz, for instance, was reported as receiving over $2.5 million from oil, gas, and coal corporations [211].

Politicians, in particular, are incredibly vulnerable to compromised funding as they have the dangerous combination of policy power and a desperate need of money. To win a U.S. election, for example, it is estimated that you need to raise nearly $50,000, every day—for six years [212]. Not too surprising then, that the former Senate Majority Leader revealed that money-gathering efforts can now make up around two-thirds of a politician's schedule [213].

And so, if an incredibly wealthy corporation were to approach a political candidate with some prerequisites and a potential multimillion-dollar 'donation', then we can see how it might appear tempting. And, as the general public cannot afford such donations, our legitimate concerns may end up at the bottom of a growing pile. Accordingly, when reviewing political funding, Princeton University concluded, "The preferences of the average American appear to have only a minuscule, near-zero, statistically non-significant impact upon public policy." [214]

Therefore, given that politicians accept billions of dollars from logging firms, cattle ranches, and fossil fuel industries, we might expect some serious conflict of interests when we ask politicians to make climate action a priority. With so much money on the other side, the public can be rendered mute in a contest of volume.

Adding to Greta's metaphor [215], the house is on fire, yet those who profit from selling the fuel can set the priorities.

As mentioned previously, in addition to political funding, it is also important to consider political focus.

The main objective of the modern-day, neoliberal politician appears to be growing the economy, as quickly as possible, regardless of the consequences. Just to clarify, in our current model, economic growth is an increase in the number of goods and services produced and sold. Prioritizing this can lead to a fundamental flaw: Overconsumption is already killing us and our environment, but capitalist politicians are hell-bent on further increasing consumption. Without stipulating sustainability. Without factoring in the hidden costs—which unfortunately includes waste, pollution, and lives.

As the goal is always more, we have systems predicated on perpetual growth. And infinite growth on a planet with finite space and finite resources is—to put it politely—exceptionally unwise.

If we consider the action items suggested thus far, they all have something in common: no longer purchasing something. And so it makes sense, given our current paradigm, that politicians have been reluctant to tackle the inconvenient truth of the climate and ecological crisis. Authentic solutions could require them to recommend actions that impede their primary focus: perpetual growth. The same authentic solutions would also require them to go against their main sources of funding: the wealthiest corporations.

We often view our political systems as incredibly inefficient, to the extent that it appears as though nothing is happening. But it is far worse than nothing happening. Our political systems often have us head-down, running at record-breaking speeds in the complete opposite direction of where we need to be going. While the world's top scientists are crying out that we need to cut emissions to net zero, our politicians are aiming to triple the global economy in just a few decades [216]—which would equate to rapidly increasing production, and therefore, rapidly increasing emissions.

Many of the symptoms we are experiencing, from physical and mental health epidemics to the destruction of our ecosystems, are deeply connected, and stem from this fundamental systemic error: the modus operandi on our finite planet is stuck at more power, more profit, and more consumption. There is no logical endpoint when the target is more.

In the late '80s, leading scientists clearly warned our politicians that not reducing emissions will harm and kill future generations [217] and yet, they chose to increase emissions: politics-as-usual. The largest fossil fuel companies have also known for decades that their practices are harming us and the natural world—their own scientists informed them [218]. In response, they funded disinformation campaigns [219], and expanded their operations: business-as-usual.

Decades later, emissions are continuing to rise, the biggest corporations are only acting concerned, and our politicians are still biding their time. Therefore, some form of system change is required for survival. Not incremental tweaks in line with special interest groups, but drastic, meaningful change in line with the united science. We need to end unsustainable systems that eat the planet to feed privatized profits.

And so, what can we do about this? How can we prevent unethical corporate funding? How can we redirect political focus? How can we change the system?

To tackle the funding issue we can attack it from both sides, starting by no longer funding unethical corporations. While we might look at the most polluting organizations as too big to fail, we forget that their money was once our money. The reason they continue to operate is because we continue to pay their bills. We continue to buy their products. By no longer purchasing unethical and unsustainable products, we cut off their funding at the source, and we seriously limit their ability to influence politicians.

And attacking it from the other side, we can stop voting for compromised political candidates. If we vote for candidates who are not funded by the largest corporations but are instead funded by the people, then they are far more likely to prioritize our legitimate concerns.

To help shift political focus, there are many things we can do. First and foremost, we can ask our politicians and representatives to declare a climate emergency. While this might seem incredibly ambitious, after public pressure, the UK and Ireland officially declared a climate emergency [220]. And since then, other nations have followed suit [221]. Yet again, thanks to strong and persistent public persuasion.

Once our politicians acknowledge the severity of the issue and declare an emergency, we need to be sure they treat it as such and that they take swift and decisive action. We need to ensure that they transition to sustainable energy as soon as possible [222], while rapidly rebuilding and expanding natural ecosystems so we can cut emissions and pay back our carbon debt. Put simply, we need to push them to reach the most ambitious targets within the Paris Agreement [223].

However, we will have to remain vigilant because corporate PR and empty political promises are far more familiar, and significantly easier, than the required level of corporate regulation and pioneering policies. For example, only 24 hours after the Canadian government declared a climate emergency [224], they announced a multibillion-dollar investment into a vast fossil fuel expansion project [225]. In other words, the green talk is far easier than the green walk.

And so, how do we persuade our politicians to declare an emergency and act urgently?

We ask them. A lot.

We can barrage them with letters, phone calls, and social media messages. And, given that time is against us, and people are suffering, I'd also encourage you to partake in the next global climate strike.

Fortunately, when it comes to system change, we can discuss a positive tipping point as history reveals that a modest maximum of only 3.5% of the population [226] is what it takes for a peaceful movement to succeed (details of the next global climate strike are always posted on my Instagram page).

I haven't given up on team human. I am confident that if we communicate peacefully and respectfully, with clear, direct goals, backed by science, and if we are relentless, then we will achieve many great things within our lifetimes.

ACTION ITEM No 4:

CHANGE THE SYSTEM

* Stop funding unethical corporations, stop voting for compromised politicians, demand that politicians declare a climate emergency, and pressure them to meet the more ambitious targets of the Paris Agreement as soon as possible.

Many have argued that system change is the only solution. Others have argued that personal change is the only solution. I argue that both are required, and both complement each other.

Politicians will often base policies around what they predict will obtain the most votes, and if they see the public making meaningful lifestyle changes to combat the climate crisis, they will see that this is incredibly important to us. Therefore, to win our votes, they will propose more meaningful, sustainably-minded policies.

The Paris Agreement highlights that consumption and lifestyle changes play an important role in combating the climate crisis [227]. And as we are asking our politicians to follow the agreement, we too must play our part. We must make significant behavioral changes. We must change life-as-usual.

As it stands, there is a global bystander-esque effect where most are aware of the crisis, but since most are still carrying on with their existing behaviour, many are uninspired to act or take it seriously. However, when we act as though we are in an emergency, we evoke urgency from those around us. It is our action that truly displays our values and priorities. By walking the walk, we can speak with integrity and so more people will listen, and more people will act.

What's more, getting involved in more proactive solutions is a known way to combat the anxiety that can come with increased climate awareness.

In short, don't wait for others to act. Instead, be first to be kind. The accumulative effect of this mentality is necessary, mighty, and highly contagious.

Change is coming. Dare to be one of the pioneers.

CHAPTER 7

UPGRADES

If we are still for a moment and we take a look around, we can find an array of unsettling scenes; we can see that if we were to scientifically redesign society from scratch, it wouldn't look like this. When we see that the majority of items sold are bad for us and/or our environment, when we see the streams of cars that dominate our cities and the seas of faces staring into screens, when we start to notice our own mental and physical health deteriorate from artificial and sedentary lifestyles, and when we see that there is much more to life than the daily grind—these scenes remind us that it is high time to upgrade.

It seems that flawed incentives drove us into the industrial age, and sheer momentum pushed us into the digital age. And now, before we dive headfirst into an age of automation, we ought to take a detour via a much-needed age of critical reflection.

We need to really reflect on why we do the things we do. And rather than seeing where we end up as we are carried along by the current of perpetual growth, we can design a deeply considered vision for the future.

Rather than making decisions based on what makes the most business sense, we can think with regards to what makes scientific sense for the environment and our own good health. The beauty of this is that a smarter future is far from a dull and grey one—it is a future with clean air, clean oceans, and abundant, sustainable ecological systems, where we don't deny the science or allow harmful systems to perpetuate.

We can design a future with less superficial, transient, material consumption, more authentic connections, and improved environmental, mental, and physical health. We can manifest a future where we don't have to be ashamed of our systems and lifestyles. We have all of the ingredients for a massive upgrade, we just need to re-evaluate, reprioritize, and reorganize. Paradise is earth, it is just currently buried under a thin layer of short-sighted, man-made bullshit.

If we are to reach our full potential and be worthy of the title of Homo sapiens—wise beings—we will need to continue to examine the status quo. We will need to reconsider many things that have gradually become the new norm:

- offshoring wealth and offshoring emissions
- extraction-based linear production chains
- resource and information monopolies
- dependant fractured communities
- the military-industrial complex
- symptom treatments
- privatized commons
- theatre democracy
- meaningless jobs
- and so on

We will need to change what we value and how we measure progress. We will need to continue to question our systems, lifestyles, tools, ideas, and stories. We will need to continue to set new standards as even the most innovative vision that we can muster will likely carry some residue from our own compromised lens. After all, us Homo sapiens are but temporary conglomerates of limited perspectives and limited experiences.

We cannot afford to become married to our ideas or become inflexible. Compassion, evidence, and a great dose of humility can keep us on the right track, away from dead-ends and dogma. And so, we shouldn't aim for a static destination, but an ever-evolving, self-criticizing, system of science.

CHAPTER 8

GENERATIONS

We often look back through history and critique previous behaviours, but how will future generations look back on us?

Perhaps they won't understand why we consumed well beyond sustainable limits and why we didn't treat the climate emergency as an actual emergency. They might wonder why we didn’t demand action from our complicit politicians and why we continued to fund corporations that were destroying our environment. They might wish we had acted during the make or break years before we set off a series of fatal amplifying feedback loops. They might wonder why we adopted—what is perhaps the most limiting and dangerous mindset of all—the self-fulfilling mindset that we cannot personally make a difference. They might look back in utter disbelief, with great anger and frustration, as while we might blame our circumstances, they may blame our inaction.

Alternatively, future generations may look back with sympathy, admiration, and pride. They might say that we were a generation that faced a difficult challenge: inheriting certain systems, traditions, and lifestyles, but then becoming increasingly aware of the harmful consequences. They might sympathize that it must have been difficult to self-criticize and break deeply ingrained modes of behavior within a short timeframe. Perhaps our grandchildren will admire that we acknowledged the error of our ways and embraced the challenge of change. They might respect how we didn't allow ourselves to become passive bystanders who accepted defeat and complied with harmful trends.

Future generations may call our era the ethical renaissance, one of mass awakening where, instead of living day-to-day on autopilot, we deeply considered the complicated repercussions of our actions; where we shattered the illusion of individualism and realized that what we do or don't do has significant consequences for those around us, the environment, and generations to come; where we acknowledged that we aren't an array of atomized, insignificant individuals in standalone bubbles, but are instead a highly social species that form part of a complicated web of interconnections far beyond our own limited perspective.

Although it may seem like a freight train of chaos is hurtling towards us, we are not yet tied to the tracks—we can abandon the echo of old narratives.

We will need to remain cautious in these strange times as it is far easier to justify current choices than depart from our comfort zones. When we fear change we often quickly create stories that can justify inaction—often without knowing, often by default.

True wisdom is not acquired by simply capturing knowledge; it is earned by acting upon knowledge. And true bravery is not the absence of fear, it is doing what you feel is the right thing to do—despite the fear.

Acting against our better intuition, and opting for the comfortable options, often results in the most regret. Instead, we could become a generation that had the wisdom and bravery to tackle highly inconvenient and disruptive truths, a generation who acknowledged that inaction results in greater human suffering and understood that compassion without action is merely observation.

We could become a generation that not only wanted change but were willing to change. A generation not only against the climate crisis but proactively for the available solutions. A generation who were not set in their ways but got better with age. Who lead by example, not opinion.

We don't have to drag the past with us; we can break the mould. We can decide what generation we will become. We can choose to embrace the opportunity to be better.

CHAPTER 9

INITIATE

What greater cause can we fight for than the protection of our own home planet and the raw ingredients of life?

We need to declare war on the climate and ecological crisis. A war with genuine boots-on-the-ground determined action, where we rise up to defend our communities—with wartime focus and wartime efforts. A very unique war where every citizen, from every nation, can fight a common enemy—where instead of ending lives, we have the absolute privilege to protect the lives of our children and their children.

Currently, we are losing the battle, but we cannot afford to surrender. We can prove the bunker-building cynics wrong. We can show them that us Homo sapiens are not incapable of sacrificing convenience to protect future generations. We can show them, that rather than defending unrestrained corporatism till the bitter end, we can instead defend our neighbours and the natural world. We can show the cynics that before it's too late, us Homo sapiens, as flawed as we may currently be, will stand up locally, nationally, and internationally, and place planet and people before pollutants and profits.

With fighting spirit, we can rise above the dominant harmful ideas of our time and be part of something incredibly powerful and positive—something we can be proud of. We can each accept our moral duty as citizens of this beautiful planet—a planet we were gifted. We can do the right thing, we can choose to be kind, we can claim our place on the right side of history.

This is not a drill. And there is no planet B. The time for excuses, speculation, or middle ground compromises is over. It is overwhelmingly evident that there is a genuine fatal emergency and we need to act right now—future generations are counting on it. We need to rapidly decrease consumption and emissions. We need to make sustainability a priority—as a matter of survival. We need to accept this mission.

We have had decades of prep talk, now we must embrace the solutions. We must answer the call and ascend to the defining challenge of our era. This is zero hour. It is time to initiate Operation Sustainable Human.

KEY FACTS

- HUMANS ARE CAUSING GLOBAL WARMING
- POLLUTION SERIOUSLY HARMS AND KILLS
- RAPID DRASTIC CHANGE IS REQUIRED

ACTION ITEMS

- SWITCH TO A PLANT-BASED DIET
- AVOID FOSSIL FUEL TRAVEL
- CURB YOUR CONSUMPTION
- CHANGE THE SYSTEM

In addition to the action items, please help to share the project. This book was created by volunteer scientists, and so any help spreading the message will be hugely appreciated.

If you have any questions or are looking for further resources, please don't hesitate to get in contact.

I hope this book has been helpful.

Thank you.

References

1 **UN. 2015.** Historic paris agreement on climate change. *The United Nations, Climate Change* Dec 15
unfccc.int/news/finale-cop21

2 **Singer C, McCarthy J, Sanchez E. 2018.** Countries of the paris climate agreement. *Global Citizen* Oct 12
globalcitizen.org/en/content/7-countries-paris-climate-agreeement

3 **UN. 2015.** Paris Agreement. *The United Nations* Dec 12
unfccc.int/sites/default/files/english_paris_agreement.pdf

4 **Cook J, Nuccitelli D, Green SA, Richardson M, Winkler B, Painting R, Way R, Jacobs P, Skuce A. 2013.** Quantifying the consensus on anthropogenic global warming in the scientific literature. *Environmental Research Letters* 8:(2)
opscience.iop.org/article/10.1088/1748-9326/8/2/024024/meta

5 **Cook J, Oreskes N, Doran PT, Anderegg WRL, Verheggen B, Maibach EW, Carlton JS, Lewandowsky S, Skuce AG, Green SA, Nuccitelli D, Jacobs P, Richardson M, Winkler B, Painting R, Rice K. 2016.** Consensus on consensus: a synthesis of consensus estimates on human-caused global warming. *Environmental Research Letters* 11:(4)
doi.org/10.1088/1748-9326/11/4/048002

6 **EDG. 2019.** Climate facts. *Environmental Defence Fund* May 9
edf.org/climate/how-climate-change-plunders-planet/climate-facts

7 **Scientific American. 2019.** *Are Volcanoes or Humans Harder on the Atmosphere?*
scientificamerican.com/article/earthtalks-volcanoes-or-humans

8 **EPA. 2019.** Causes of climate change. *Environmental Protection Agency* May 7
19january2017snapshot.epa.gov/climate-change-science/causes-climate-change_.html

9 **Jiang XQ, Mei XD, Feng D. 2016.** Air pollution and chronic airway diseases: what should people know and do? *Journal of thoracic disease* 8(1): E31–E40
doi.org/10.3978/j.issn.2072-1439.2015.11.50

10 **Esposito S, Galeone C, Lelii M, Longhi B, Ascolese B, Senatore L, Prada E, Montinaro V, Malerba S, Patria MF, Principi N. 2014.** Impact of air pollution on respiratory diseases in children with recurrent wheezing or asthma. *BMC pulmonary medicine* 14 130

doi.org/10.1186/1471-2466-14-130

11 **Kim D, Chen Z, Zhou LF, Huang SX. 2018.** Air pollutants and early origins of respiratory diseases. *Chronic Diseases and Traditional Medicine* (4):2 75-95
doi.org/10.1016/j.cdtm.2018.03.003

12 **Mason SA, Welch V, Neratko J. 2019.** Synthetic polymer contamination in bottled water. *State University of New York at Fredonia, Department of Geology & Environmental Sciences* Apr 29
orbmedia.org/sites/default/files/FinalBottledWaterReport.pdf

13 **Mason SA, Welch V, Neratko J. 2019.** Synthetic polymer contamination in bottled water. *State University of New York at Fredonia, Department of Geology & Environmental Sciences* Apr 29
orbmedia.org/sites/default/files/FinalBottledWaterReport.pdf

14 **Wieczorek AM, Morrison L, Croot PL, Allcock AL, MacLoughlin E, Savard O, Brownlow H and Doyle TK. 2018.** Frequency of Microplastics in Mesopelagic Fishes from the Northwest Atlantic. *Frontiers Sci* 5:39
doi.org/10.3389/fmars.2018.00039

15 **Wilcox C, Sebille EV, Hardesty BD. 2015.** Plastic in seabirds is pervasive and increasing,
Proceedings of the National Academy of Sciences 201502108
doi.org/10.1073/pnas.1502108112

16 **Smith M, Love DC, Rochman CM, Neff RA. 2018.** Microplastics in Seafood and the Implications for Human Health. *Current environmental health reports* 5:(3) 375–386
doi.org/10.1007/s40572-018-0206-z

17 **Sciammacco S. 2014.** Yoga mat chemical found in nearly 500 foods. *EWG* Feb 27
ewg.org/release/yoga-mat-chemical-found-nearly-500-foods#.W3xeHNhKgWo

18 **UOA. 2015.** Exposure to harmful phthalates from processed foods and soft drinks. *University of Adelaide* Apr 16
phys.org/news/2015-04-exposure-phthalates-foods-soft.html

19 **Cox KD, Covernton GA, Davies HL, Dower JF, Juanes F, Dudas SE. 2019.** Human Consumption of Microplastics. *Environ Sci Technol* 53:(12) 7068-7074
doi.org/10.1021/acs.est.9b01517

20 **Dell'Amore C. 2010.** Chemical BPA Linked to Heart Disease, Study Confirms. *National Geographic* Jan 17 news.nationalgeographic.com/news/2010/01/100115-bpa-bisphenol-a-heart-disease

21 **Cox KD, Covernton GA, Davies HL, Dower JF, Juanes F, Dudas SE. 2019.** Human Consumption of Microplastics. *Environ Sci Technol* 53:(12) 7068-7074 doi.org/10.1021/acs.est.9b01517

22 **UON. 2019.** Plastic ingestion by people could be equating to a credit card a week. *University of Newcastle* Jun 12 newcastle.edu.au/newsroom/featured/plastic-ingestion-by-people-could-be-equating-to-a-credit-card-a-week

23 **Kim JS, Lee HJ, Kim SK, Kim HJ. 2018.** Global pattern of microplastics in commercial grade salts: sea salt as an indicator of seawater microplastic pollution. *Environ Sci Technol* 52: 12819–12828 doi.org/10.1021/acs.est.8b04180

24 **Dell'Amore C. 2010.** Chemical BPA Linked to Heart Disease, Study Confirms. *National Geographic* Jan 17 news.nationalgeographic.com/news/2010/01/100115-bpa-bisphenol-a-heart-disease

25 **Dell'Amore C. 2010.** Chemical BPA Linked to Heart Disease, Study Confirms. *National Geographic* Jan 17 news.nationalgeographic.com/news/2010/01/100115-bpa-bisphenol-a-heart-disease

26 **Galloway TS, Baglin N, Lee BP, Kocur AL, Shepherd MH, Steele AM, BPA Schools Study Consortium. 2018.** An engaged research study to assess the effect of a 'real-world' dietary intervention on urinary bisphenol A (BPA) levels in teenagers. *BMJ Open* 2018;8:e018742 doi.org/10.1136/bmjopen-2017-018742

27 **Dell'Amore C. 2010.** Chemical BPA Linked to Heart Disease, Study Confirms. *National Geographic* Jan 17 news.nationalgeographic.com/news/2010/01/100115-bpa-bisphenol-a-heart-disease

28 **Konieczna A, Rutkowska A, Rachoń D. 2015.** Health risk of exposure to bisphenol a (BPA). *Rocz Panstw Zakl Hig* 66:(1) 5–11 ncbi.nlm.nih.gov/pubmed/25813067

29 **Olchowik-Grabarek E, Makarova K, Mavlyanov S, Abdullajanova N, Zamaraeva M. 2018.** Comparative analysis of BPA and HQ toxic impacts on human erythrocytes, protective effect mechanism of tannins (Rhus typhina). *Environ Sci Pollut Res* 25:(2) 1200-1209
doi.org/10.1007/s11356-017-0520-2

30 **WHO. 2019.** Air pollution. *The World Health Organization* Jun 12
who.int/airpollution/en

31 **Vidal J. 2009.** Global warming causes 300,000 deaths a year. *The Guardian Environment* May 29
theguardian.com/environment/2009/may/29/1

32 **Chestney N. 2012.** 100 mln will die by 2030 if world fails to act on climate. *Reuters* Sep 25
in.reuters.com/article/climate-inaction/100-mln-will-die-by-2030-if-world-fails-to-act-on-climate-report-idINDEE88O0HH20120925

33 **Dockrill P. 2019.** We're headed for 'climate apartheid'. *Business Insider* Jun 26
businessinsider.com/climate-apartheid-united-nations-report-2019-6

34 **Cosier S. 2019.** Get ready for tens of millions of climate refugees. *MIT* Apr 24
technologyreview.com/s/613342/get-ready-for-tens-of-millions-of-climate-refugees

35 **Taylor M. 2017.** Climate change will create world's biggest refugee crisis. *The Guardian* Nov 2
theguardian.com/environment/2017/nov/02/climate-change-will-create-worlds-biggest-refugee-crisis

36 **Haines A, Ebi K. 2019.** The Imperative for Climate Action to Protect Health. *New England Medical Journal* 380: 263-273
doi.org/10.1056/NEJMra1807873

37 **Christensen J. 2019.** 250,000 deaths a year from climate change is a 'conservative estimate'. *CNN* Jan 16
cnn.com/2019/01/16/health/climate-change-health-emergency-study/index.html

38 **Vidal J. 2009.** Global warming causes 300,000 deaths a year. *The Guardian Environment* May 29
theguardian.com/environment/2009/may/29/1

39 **NG. 2019.** Health risks. *National Geographic* May 7
nationalgeographic.com/climate-change/how-to-live-with-it/health.html

40 **Thunberg G. 2019.** 'Our house is on fire': Greta Thunberg, 16, urges leaders to act on climate, *The Guardian Climate* Jan 25
theguardian.com/environment/2019/jan/25/our-house-is-on-fire-greta-thunberg16-urges-leaders-to-act-on-climate

41 **Carrington D. 2017.** Earth's sixth mass extinction event is under way. *The Guardian* Jul 17
theguardian.com/environment/2017/jul/10/earths-sixth-mass-extinction-event-already-underway-scientists-warn

42 **Woodward A. 2019.** 17 signs we're in the middle of a 6th mass extinction. *Business Insider* May 6
businessinsider.com/signs-of-6th-mass-extinction-2019-3

43 **Vidal J. 2010.** UN environment programme: 200 species extinct every day. *The Huffington Post* Aug 17
huffingtonpost.ca/2010/08/17/un-environment-programme-_n_684562.html

44 **Bradshaw CJA. 2012.** Little left to lose: deforestation and forest degradation in Australia since European colonization, *Journal of Plant Ecology* 5:(1) 109–120
doi.org/10.1093/jpe/rtr038

45 **The Rainforest Alliance. 2019.** Worried about deforestation? *Rainforest Alliance* Apr 30
my.rainforest-alliance.org/site/PageServer?pagename=issues_forest

46 **Vidal J. 2017.** We are destroying rainforests so quickly they may be gone in 100 years. *The Guardian* Climate Jan 23
theguardian.com/global-development-professionals-network/2017/jan/23/destroying-rainforests-quickly-gone-100-years-deforestation

47 **Sanchez E, Sepehr J, McCarthy J. 2018.** Humanity has killed 83% of all wild mammals. *Global Citizen* May 22
globalcitizen.org/en/content/humans-destroyed-83-of-wildlife-report

48 **SA. 2015.** Why should we help conserve wildlife? *Share America* Mar 2
share.america.gov/why-should-we-help-conserve-wildlife

49 **Carrington D. 2018.** Humans are just 0.01% of all life but have destroyed 83% of wild mammals. *The Guardian* May 21 theguardian.com/environment/2018/may/21/human-race-just-001-of-all-life-but-has-destroyed-over-80-of-wild-mammals-study

50 **Vidal J. 2010.** UN environment programme: 200 species extinct every day. *The Huffington Post* Aug 17 huffingtonpost.ca/2010/08/17/un-environment-programme-_n_684562.html

51 **NWF. 2019.** The science of climate change. *National Wildlife Federation* May 9 nwf.org/Eco-Schools-USA/Become-an-Eco-School/Pathways/Climate-Change/Facts

52 **Witschge L. 2018.** Why are coral reefs import, and why are they dying? *Aljazeera* Jan 29 aljazeera.com/indepth/features/coral-reefs-important-dying-180128135520949.html

53 **WWF. 2015.** Living blue planet report: species, habitats and human well-being. *World Wildlife Foundation* ocean.panda.org/media/Living_Blue_Planet_Report_2015_Final_LR.pdf

54 **Arsenault C. 2019.** Only 60 Years of Farming Left If Soil Degradation Continues, *Scientific American* Apr 30 scientificamerican.com/article/only-60-years-of-farming-left-if-soil-degradation-continues

55 **WEF. 2012.** What if the world's soil runs out? *World Economic Forum* Dec 14 world.time.com/2012/12/14/what-if-the-worlds-soil-runs-out

56 **Roach J. 2006.** Seafood may be gone by 2048, *National Geographic* Nov 2 nationalgeographic.com/animals/2006/11/seafood-biodiversity

57 **Clover C. 2006.** All seafood will run out in 2050. *Telegraph* Nov 3 telegraph.co.uk/news/uknews/1533125/All-seafood-will-run-out-in-2050-say-scientists.html

58 **Tidey A. 2019.** Arctic could be ice-free in the summer from as early as 2030. *Euro News* May 3

euronews.com/2019/03/05/arctic-could-be-ice-free-in-the-summer-from-as-early-as-2030-study

59 **Johnston I. 2016.** Arctic could become ice-free for first time in more than 100,000 years. *The Independent* Jun 4 independent.co.uk/environment/climate-change/arctic-could-become-ice-free-for-first-time-in-more-than-100000-years-claims-leading-scientist-a7065781.html

60 **Pearce F. 1989.** Surge in carbon dioxide prompts new greenhouse fears. *New Scientist* Apr 1 newscientist.com/article/mg12216581-800-surge-in-carbon-dioxide-prompts-new-greenhouse-fears

61 **Jones N. 2017.** How the world passed a carbon threshold and why it matters. *Yale Environment 360* Jan 26 e360.yale.edu/features/how-the-world-passed-a-carbon-threshold-400ppm-and-why-it-matters

62 **Science Daily. 2019.** Warming Arctic permafrost releasing large amounts of potent greenhouse gas. *Science Daily* Apr 15 sciencedaily.com/releases/2019/04/190415090848.htm

63 **Wilkerson J, Dobosy R, Sayres DS, Healy C, Dumas E, Baker B, Anderson JG. 2019.** Permafrost nitrous oxide emissions observed on a landscape scale using the airborne eddy-covariance method. *Atmospheric Chemistry and Physics* 19:(7) 4257 doi.org/10.5194/acp-19-4257-2019

64 **Perrone A. 2019.** Climate change: Arctic permafrost now melting at levels not expected until 2090. *The Independent* Jun 15 independent.co.uk/news/world/americas/climate-change-breakdown-arctic-frost-thawing-canada-environment-a8959056.html

65 **Green M. 2019.** Scientists amazed as Canadian permafrost thaws 70 years early. *Reuters* Jun 18 reuters.com/article/us-climate-change-permafrost/scientists-amazed-as-canadian-permafrost-thaws-70-years-early-iduskcn1tj1xn

66 **EPA. 2019.** Causes of climate change. *Environmental Protection Agency* May 7 19january2017snapshot.epa.gov/climate-change-science/causes-climate-change_.html

67 **Cook J, Nuccitelli D, Green SA, Richardson M, Winkler B, Painting R, Way R, Jacobs P, Skuce A. 2013.** Quantifying the

consensus on anthropogenic global warming in the scientific literature. *Environmental Research Letters* 8:(2)
opscience.iop.org/article/10.1088/1748-9326/8/2/024024/meta

68 **McKibben B. 2019.** Notes from a remarkable political moment for climate change. *The New Yorker* May 1
newyorker.com/news/daily-comment/notes-from-a-remarkable-political-moment-for-climate-change

69 **Cook J, Oreskes N, Doran PT, Anderegg WRL, Verheggen B, Maibach EW, Carlton JS, Lewandowsky S, Skuce AG, Green SA, Nuccitelli D, Jacobs P, Richardson M, Winkler B, Painting R, Rice K. 2016.** Consensus on consensus: a synthesis of consensus estimates on human-caused global warming. *Environmental Research Letters* 11:(4)
doi.org/10.1088/1748-9326/11/4/048002

70 **Mortillaro N. 2017.** More than 15,000 scientists from 184 countries issue warning to humanity, *CBC News* Nov 13
cbc.ca/news/technology/15000-scientists-warning-to-humanity-1.4395767

71 **Watts J. 2018.** We have 12 years to limit climate change catastrophe, warns UN, *The Guardian Climate* Oct 8
theguardian.com/environment/2018/oct/08/global-warming-must-not-exceed-15c-warns-landmark-un-report

72 **Dockrill P. 2019.** We're headed for 'climate apartheid'. *Business Insider* Jun 26
businessinsider.com/climate-apartheid-united-nations-report-2019-6

73 **Eshel G, Shepon A, Makov T, Milo R. 2014.** Environmental costs of animal-based categories. Proceedings of the National Academy of Sciences 111:(33) 11996-12001
doi.org/10.1073/pnas.1402183111

74 **Koneswaran G, Nierenberg D. 2008.** Global farm animal production and global warming: impacting and mitigating climate change. *Environ Health Perspect* 116(5): 578–582
doi.org/10.1289/ehp.11034

75 **Rainforest Foundation. 2019.** Effects of Agriculture. *Rainforest Foundation* May 2
rainforestfoundation.org/agriculture

76 **Butler R. 2017.** Rainforest Facts. *Mongabay* Jan
rainforests.mongabay.com/facts/rainforest-facts.html#8

77 **Scientific American. 2019.** Measuring the Daily Destruction of the World's Rainforests. *Scientific American* May 2
scientificamerican.com/article/earth-talks-daily-destruction

78 **Butler R. 2012.** Cattle Ranching's Impact on the Rainforest. *Mongabay* July
rainforests.mongabay.com/0812.htm

79 **Veiga JB, Tourrand JF, Poccard-Chapuis R, Piketty MG. 2003.** Cattle Ranching in the Amazon Rainforest. *World Forestry Congress* 0568-B1
fao.org/3/XII/0568-B1.htm

80 **Yale. 2019.** Soy Agriculture in the Amazon Basin. *Yale School of Forestry and Environmental Studies: Global Forest Atlas* May 2
globalforestatlas.yale.edu/amazon/land-use/soy

81 **Koneswaran G, Nierenberg D. 2008.** Global farm animal production and global warming: impacting and mitigating climate change. *Environ Health Perspect* 116(5): 578–582
doi.org/10.1289/ehp.11034

82 **Walsh B. 2013.** The triple whopper environmental impact of global meat production. *Time* Dec 16
science.time.com/2013/12/16/the-triple-whopper-environmental-impact-of-global-meat-production

83 **UN. 2018.** Tackling the world's most urgent problem: meat. *The United Nations* Sep 26
unenvironment.org/news-and-stories/story/tackling-worlds-most-urgent-problem-meat

84 **Campbell TC.** 2019. Food Footnotes. *The Plantrician Project* Jun 3
plantricianproject.org/footnotes

85 **Peta. 2019.** Meat and the environment. *Peta* May 2
peta.org/issues/animals-used-for-food/meat-environment

86 **Peta. 2019.** Meat and the environment. *Peta* May 2
peta.org/issues/animals-used-for-food/meat-environment

87 **Pimentel D, Berger B, Filiberto D, Newton M, Wolfe B,**

Karabinakis E, Clark S, Poon E, Abbett E, Nandagopal S. 2004. Water Resources: Agricultural and Environmental Issues. *BioScience* 54:(10) 909-918
doi.org/10.1641/0006-3568(2004)054[0909:WRAAEI]2.0.CO;2

88 **Peta. 2019.** Meat and the environment. *Peta* May 2
peta.org/issues/animals-used-for-food/meat-environment

89 **EPA. 2004.** Risk Assessment Evaluation for Concentrated Animal Feeding Operations. *Environmental Protection Agency* May
nepis.epa.gov/Exe/ (search 600R04042)

90 **AUM Media. 2019.** The facts. *AUM* May 2
cowspiracy.com/facts

91 **Carrington D. 2014.** Giving up beef will reduce carbon footprint more than cars. *The Guardian Climate* Jul 21
theguardian.com/environment/2014/jul/21/giving-up-beef-reduce-carbon-footprint-more-than-cars

92 **France-Press A. 2017.** Methane emissions from cattle are 11% higher than estimated. *The Guardian Climate* Sep 29
theguardian.com/environment/2017/sep/29/methane-emissions-cattle-11-percent-higher-than-estimated

93 **Johnson KA, Johnson DE. 1995.** Methane emissions from cattle. *J Anim Sci* 73:(8) 2483-2492
ncbi.nlm.nih.gov/pubmed/8567486

94 **France-Press A. 2017.** Methane emissions from cattle are 11% higher than estimated. *The Guardian Climate* Sep 29
theguardian.com/environment/2017/sep/29/methane-emissions-cattle-11-percent-higher-than-estimated

95 **Shindell DT, Faluvegi G, Koch DM, Schmidt GA, Unger N, Bauer SE. 2009.** Improved attribution of climate forcing to emissions. *Science* Oct 30 326:(5953) 716-718
ncbi.nlm.nih.gov/pubmed/19900930

96 **United Nations. 2006.** Livestock's Long Shadow: Environmental issues and options. *Food and agriculture of the united nations*
fao.org/3/a0701e/a0701e00.htm

97 **United Nations. 2006.** Livestock's Long Shadow: Environmental issues and options. *Food and agriculture of the united nations*
fao.org/3/a0701e/a0701e00.htm

98 **Hamblin J. 2017.** If everyone ate beans instead of beef. *The Atlantic* Aug 2
theatlantic.com/health/archive/2017/08/if-everyone-ate-beans-instead-of-beef/535536/

99 **UN. 2015.** Paris Agreement. *The United Nations* Dec 12
unfccc.int/sites/default/files/english_paris_agreement.pdf

100 **Hamblin J. 2017.** If everyone ate beans instead of beef. *The Atlantic* Aug 2
theatlantic.com/health/archive/2017/08/if-everyone-ate-beans-instead-of-beef/535536/

101 **Edsor B. 2017.** These 14 elite athletes are vegan. *The Business Insider* Nov 1
businessinsider.com/elite-athletes-who-are-vegan-and-what-made-them-switch-their-diet-2017-10

102 **Peta. 2018.** 6 Olympic champions you probably had no clue were vegan. *Peta* Feb 21
peta.org/blog/6-olympic-champions-probably-no-clue-vegan

103 **Andrade M. 2017.** 15 Seriously shredded vegan bodybuilders. *Mens Health* Sep 20
menshealth.com/fitness/a19535559/vegan-bodybuilders-instagram

104 **Peta. 2014.** 6 Runners who put the victory in vegan. *Peta* Jul 17
peta.org/living/food/vegan-runners

105 **AUM Media. 2019.** The facts. *AUM* May 2
cowspiracy.com/facts

106 **Moskin J, Plumer B, Lieberman R, Weingart E. 2019.** Your questions about food and climate change, answered. *The New York Times* Apr 30
nytimes.com/interactive/2019/04/30/dining/climate-change-food-eating-habits.html

107 **Loria J. 2017.** 15000 Scientists from 184 countries urge the people to go vegan. *MFA* Nov 17
mercyforanimals.org/15000-scientists-from-184-countries-urge

108 **TVC. 2019.** How much have you saved? *TVC* May 4
thevegancalculator.com/#calculator

109 **WAF. 2019.** Animal agriculture causing animal extinction. *The World Animal Foundation* May 4 worldanimalfoundation.org/articles/article/8949042/186425.htm

110 **Cameron J, Cameron SA. 2017.** Animal agriculture is choking the earth and making us sick. *Th Guardian* Dec 4 theguardian.com/commentisfree/2017/dec/04/animal-agriculture-choking-earth-making-sick-climate-food-environmental-impact-james-cameron-suzy-amis-cameron

111 **Scientific American. 2019.** What causes ocean dead zones. *Scientific American* May 4 scientificamerican.com/article/ocean-dead-zones

112 **WAF. 2019.** Animal agriculture causing animal extinction. *The World Animal Foundation* May 4 worldanimalfoundation.org/articles/article/8949042/186425.htm

113 **Schauber G. 2017.** Cows, pigs, and poultry. *UBC* Feb 7 blogs.ubc.ca/makingwaves/2017/02/07/cows-pigs-and-poultry-the-leading-cause-of-ocean-dead-zones

114 **United Nations. 2006.** Livestock's Long Shadow: Environmental issues and options. *Food and agriculture of the united nations* fao.org/3/a0701e/a0701e00.htm

115 **WAF. 2019.** Animal agriculture causing animal extinction. *The World Animal Foundation* May 4 worldanimalfoundation.org/articles/article/8949042/186425.htm

116 **Hansel H. 2018.** How animal agriculture affects our planet. *Pachamama Alliance* Feb 2 blog.pachamama.org/how-animal-agriculture-affects-our-planet

117 **WAF. 2019.** Animal agriculture causing animal extinction. *The World Animal Foundation* May 4 worldanimalfoundation.org/articles/article/8949042/186425.htm

118 **USGS. 2006.** Pesticides in the Nation's Streams and Ground Water. *USGS* Mar pubs.usgs.gov/fs/2006/3028/

119 **Koneswaran G, Nierenberg D. 2008.** Global farm animal production and global warming: impacting and mitigating climate change. *Environ Health Perspect* 116(5): 578–582 doi.org/10.1289/ehp.11034

120 **United Nations. 2006.** Livestock's Long Shadow: Environmental issues and options. *Food and agriculture of the united nations* fao.org/3/a0701e/a0701e00.htm

121 **Koneswaran G, Nierenberg D. 2008.** Global farm animal production and global warming: impacting and mitigating climate change. *Environ Health Perspect* 116(5): 578–582 doi.org/10.1289/ehp.11034

122 **Carrington D. 2018.** Avoiding meat and dairy is 'single biggest way' to reduce your impact on Earth. *The Guardian Climate* May 31 theguardian.com/environment/2018/may/31/avoiding-meat-and-dairy-is-single-biggest-way-to-reduce-your-impact-on-earth

123 **Carrington D. 2018.** Global food system is broken. *The Guardian* Nov 28 theguardian.com/environment/2018/nov/28/global-food-system-is-broken-say-worlds-science-academies

124 **EcoWatch. 2018.** Going vegan is the best thing you can do for the planet. *EcoWatch* Apr 09 ecowatch.com/vegan-climate-change-2558286917.html

125 **Petter O. 2018.** Veganism is single biggest way to reduce our environmental impact. *The Independent* Jun 1 independent.co.uk/life-style/health-and-families/veganism-environmental-impact-planet-reduced-plant-based-diet-humans-study-a8378631.html

126 **Webber J. 2019.** Eating vegan is the most effective way to combat climate change, says largest-ever food production analysis. *Livekindly* May 19 livekindly.com/eating-vegan-is-the-most-effective-way-to-combat-climate-change-says-largest-ever-food-production-analysis

127 **Poore J, Nemecek T. 2019.** Reducing food's environmental impacts through producers and consumers. *Science* 360:(6392) 987-992 doi.org/10.1126/science.aaq0216

128 **Machovina B, Feeley KJ, Ripple WJ. 2015.** Biodiversity conservation: The key is reducing meat consumption. *Science of Total Environment* 536 419-431 ncbi.nlm.nih.gov/pubmed/26231772

129 **Koneswaran G, Nierenberg D. 2008.** Global farm animal

production and global warming: impacting and mitigating climate change. *Environ Health Perspect* 116(5): 578–582
doi.org/10.1289/ehp.11034

130 **EPA. 2019.** Carbon dioxide emissions. *Environmental Protection Agency* May 7
19january2017snapshot.epa.gov/ghgemissions/overview-greenhouse-gases_.html#carbon-dioxide

131 **Green J. 2019.** Effects of car pollutants on the environment. *Sciencing* Mar 13
sciencing.com/effects-car-pollutants-environment-23581.html

132 **EPA. 2019.** Green vehicle guide. *Environmental Protection Agency* May 8
epa.gov/greenvehicles/fast-facts-transportation-greenhouse-gas-emissions

133 **T&E. 2019.** CO_2 emissions from cars. *Transport and Environment* Apr 9
transportenvironment.org/publications/co2-emissions-cars-facts

134 **UN. 2019.** Get the facts. *Young Champions of the Earth, The United Nations Environment* May 8
web.unep.org/youngchampions/blog/get-facts-6-things-you-need-know-about-air-pollution

135 **Shabad LM, Smirnov GA. 1972.** Aircraft engines as a source of carcinogenic pollution in the environment. *Science Direct, Atmospheric Environment* 6:(3) 153-164
doi.org/10.1016/S0004-6981(72)80144-6

136 **Green J. 2019.** Effects of car pollutants on the environment. *Sciencing* Mar 13
sciencing.com/effects-car-pollutants-environment-23581.html

137 **Nunez C. 2019.** Air pollution, explained. *National Geographic* Feb 4
nationalgeographic.com/environment/global-warming/pollution

138 **EPA. 2019.** Basic Information about Lead Air Pollution. *Environmental Protection Agency* Apr 29
epa.gov/lead-air-pollution/basic-information-about-lead-air-pollution

139 **Ghorani-Azam A, Riahi-Zanjani B, Balali-Mood M. 2016.** Effects of air pollution on human health and practical measures for prevention in Iran. *Journal of research in medical sciences: the official journal of*

Isfahan University of Medical Sciences 21: 65
doi.org/10.4103/1735-1995.189646

140 **Green J. 2019.** Effects of car pollutants on the environment. *Sciencing* Mar 13
sciencing.com/effects-car-pollutants-environment-23581.html

141 **Carrington D. 2019.** Revealed: Air pollution may be damaging 'every organ in the body'. *The Guardian* May 17
theguardian.com/environment/ng-interactive/2019/may/17/air-pollution-may-be-damaging-every-organ-and-cell-in-the-body-finds-global-review?CMP=share_btn_link

142 **Green J. 2019.** Effects of car pollutants on the environment. *Sciencing* Mar 13
sciencing.com/effects-car-pollutants-environment-23581.html

143 **Nunez C. 2019.** Air pollution, explained. *National Geographic* Feb 4
nationalgeographic.com/environment/global-warming/pollution

144 **Green J. 2019.** Effects of car pollutants on the environment. *Sciencing* Mar 13
sciencing.com/effects-car-pollutants-environment-23581.html

145 **Nunez C. 2019.** Air pollution, explained. *National Geographic* Feb 4
nationalgeographic.com/environment/global-warming/pollution

146 **Schraufnagel DE, et al. 2019.** Air Pollution and Noncommunicable Diseases. *Chest* 155:(2) 409-416
doi.org/10.1016/j.chest.2018.10.042

147 **Mackenzie J. 2016.** Air Pollution: Everything You Need to Know. *Natural Resources Defence Council* Nov 01
nrdc.org/stories/air-pollution-everything-you-need-know#sec3

148 **Zhang X, Chen X, Zhang X. 2018.** The impact of exposure to air pollution on cognitive performance. *PNAS* 115:(37) 9193-9197
doi.org/10.1073/pnas.1809474115

149 **ECCC. 2019.** Air pollution: drivers and impacts. *Environment and Climate Change Canada* May 8
ec.gc.ca/indicateurs-indicators/default.asp?lang=En&n=D189C09D-1

150 **Carrington D. 2019.** Revealed: Air pollution may be damaging 'every organ in the body'. *The Guardian* May 17
theguardian.com/environment/ng-interactive/2019/may/17/air-

pollution-may-be-damaging-every-organ-and-cell-in-the-body-finds-global-review?CMP=share_btn_link

151 **UN. 2019.** Get the facts. *Young Champions of the Earth, The United Nations Environment* May 8 web.unep.org/youngchampions/blog/get-facts-6-things-you-need-know-about-air-pollution

152 **Carrington D. 2019.** Revealed: Air pollution may be damaging 'every organ in the body'. *The Guardian* May 17 theguardian.com/environment/ng-interactive/2019/may/17/air-pollution-may-be-damaging-every-organ-and-cell-in-the-body-finds-global-review?CMP=share_btn_link

153 **Nunez C. 2019.** Air pollution, explained. *National Geographic* Feb 4 nationalgeographic.com/environment/global-warming/pollution

154 **UN. 2019.** Get the facts. *Young Champions of the Earth, The United Nations Environment* May 8 web.unep.org/youngchampions/blog/get-facts-6-things-you-need-know-about-air-pollution

155 **McCarthy N. 2018.** Air pollution contributed to more than 6 million deaths. *Forbes* Apr 18 forbes.com/sites/niallmccarthy/2018/04/18/air-pollution-contributed-to-more-than-6-million-deaths-in-2016-infographic/#69666e0213b4

156 **Carrington D, Taylor M. 2018.** Air pollution is the 'new tobacco'. *The Guardian* Oct 27 theguardian.com/environment/2018/oct/27/air-pollution-is-the-new-tobacco-warns-who-head

157 **WWF. 2019.** Cutting aviation pollution. *World Wildlife Fund* May 9 worldwildlife.org/initiatives/cutting-aviation-pollution

158 **Schlossberg T. 2017.** Flying is bad for the planet, you can help make it better. *The New York Times* Jul 27 nytimes.com/2017/07/27/climate/airplane-pollution-global-warming.html

159 **DSF. 2017.** Air travel and climate change. *David Suzuki Foundation* Oct 5 davidsuzuki.org/what-you-can-do/air-travel-climate-change

160 **DSF. 2017.** Air travel and climate change. *David Suzuki Foundation* Oct 5

davidsuzuki.org/what-you-can-do/air-travel-climate-change

161 **Clark D. 2010.** Aviation Q&A. *The Guardian Climate* Apr 6 theguardian.com/environment/2010/apr/06/aviation-q-and-a

162 **Clark D. 2010.** The surprisingly complex truth about planes and climate change. *The Guardian Environment* Sep 9 theguardian.com/environment/blog/2010/sep/09/carbon-emissions-planes-shipping

163 **Clark D. 2010.** The surprisingly complex truth about planes and climate change. *The Guardian Environment* Sep 9 theguardian.com/environment/blog/2010/sep/09/carbon-emissions-planes-shipping

164 **Tyers R. 2017.** It's time to wake up to the devastating impact flying has on the environment. *The Conversion* Jan 11 theconversation.com/its-time-to-wake-up-to-the-devastating-impact-flying-has-on-the-environment-70953

165 **Tyers R. 2017.** It's time to wake up to the devastating impact flying has on the environment. *The Conversion* Jan 11 theconversation.com/its-time-to-wake-up-to-the-devastating-impact-flying-has-on-the-environment-70953

166 **Laville S. 2019.** Creative carbon accounting. *The Guardian Climate* Apr 25 theguardian.com/environment/2019/apr/25/uks-creative-carbon-accounting-breaches-climate-deal-say-critics

167 **Tyers R. 2017.** It's time to wake up to the devastating impact flying has on the environment. *The Conversion* Jan 11 theconversation.com/its-time-to-wake-up-to-the-devastating-impact-flying-has-on-the-environment-70953

168 **UN. 2015.** Paris Agreement. *The United Nations* Dec 12 unfccc.int/sites/default/files/english_paris_agreement.pdf

169 **GFN. 2019.** About earth overshoot day. *Global Footprint Network* Aug 2 overshootday.org/about-earth-overshoot-day

170 **Gimenez EH. 2012.** We already grow enough food for 10 billion people. *The Huffington Post* May 2 huffpost.com/entry/world-hunger_n_1463429

171 **WAP. 2019.** Animals in farming. *World Animal Protection* May 15
worldanimalprotection.org/our-work/animals-farming-supporting-70-billion-animals

172 **Duggan G. 2019.** Our fast fashion habit is killing the planet. *CBC News* Mar 10
cbc.ca/passionateeye/m_features/our-fast-fashion-habit-is-killing-the-planet

173 **Cooper KL. 2018.** Fast fashion: Inside the fight to end the silence on waste. *BBC News* Jul 31
bbc.com/news/world-44968561

174 **NCC. 2018.** The price of fast fashion. *Nature Climate Change* 8, 1
doi.org/10.1038/s41558-017-0058-9

175 **NCC. 2018.** The price of fast fashion. *Nature Climate Change* 8, 1
doi.org/10.1038/s41558-017-0058-9

176 **VEC. 2017.** Apparel textiles leverage lab. *Vancouver Economic Commission* Dec 11
vancouvereconomic.com/blog/news/apparel-textiles-leverage-lab-fast-fashion-meets-gradual-change

177 **Cooper KL. 2018.** Fast fashion: Inside the fight to end the silence on waste. *BBC News* Jul 31
bbc.com/news/world-44968561

178 **Cooper KL. 2018.** Fast fashion: Inside the fight to end the silence on waste. *BBC News* Jul 31
bbc.com/news/world-44968561

179 **Gould H. 2014.** 10 things you need to know about water impacts of the fashion industry. *The Guardian* Sep 4
theguardian.com/sustainable-business/sustainable-fashion-blog/2014/sep/04/10-things-to-know-water-impact-fashion-industry

180 **PANUK. 2019.** Pesticide concerns in cotton. *Pesticide Action Network UK* May 15
pan-uk.org/cotton

181 **Harrabin R. 2018.** Fast fashion is harming the planet. *BBC News* Oct 5
bbc.com/news/science-environment-45745242

182 **Fitzner Z. 2018.** Deforestation for fashion: The cost of rayon. *Earth*

Jun 20
earth.com/news/deforestation-fashion-rayon

183 **Perry P. 2018.** The environmental costs of fast fashion. *The Independent* Jan 8
independent.co.uk/life-style/fashion/environment-costs-fast-fashion-pollution-waste-sustainability-a8139386.html

184 **D'Alessandro N. 2014.** Facts about plastic. *Ecowatch* Apr 7
ecowatch.com/22-facts-about-plastic-pollution-and-10-things-we-can-do-about-it-1881885971.html

185 **D'Alessandro N. 2014.** Facts about plastic. *Ecowatch* Apr 7
ecowatch.com/22-facts-about-plastic-pollution-and-10-things-we-can-do-about-it-1881885971.html

186 **D'Alessandro N. 2014.** Facts about plastic. *Ecowatch* Apr 7
ecowatch.com/22-facts-about-plastic-pollution-and-10-things-we-can-do-about-it-1881885971.html

187 **Leblanc R. 2019.** The decomposition of waste in landfills. *The Balance* May 16
thebalancesmb.com/how-long-does-it-take-garbage-to-decompose-2878033

188 **Wieczorek AM, Morrison L, Croot PL, Allcock AL, MacLoughlin E, Savard O, Brownlow H and Doyle TK. 2018.** Frequency of Microplastics in Mesopelagic Fishes from the Northwest Atlantic. *Frontiers Sci* 5:39
doi.org/10.3389/fmars.2018.00039

189 **Sciammacco S. 2014.** Yoga mat chemical found in nearly 500 foods. *EWG* Feb 27
ewg.org/release/yoga-mat-chemical-found-nearly-500-foods#.W3xeHNhKgWo

190 **Mason SA, Welch V, Neratko J. 2019.** Synthetic polymer contamination in bottled water. *State University of New York at Fredonia, Department of Geology & Environmental Sciences* Apr 29
orbmedia.org/sites/default/files/FinalBottledWaterReport.pdf

191 **Dell'Amore C. 2010.** Chemical BPA Linked to Heart Disease, Study Confirms. *National Geographic* Jan 17
news.nationalgeographic.com/news/2010/01/100115-bpa-bisphenol-a-heart-disease

192 **Morelle R. 2019.** Mariana Trench: Deepest-ever sub dive. *BBC News* May 13
bbc.com/news/science-environment-48230157?fbclid=IwAR38wFJZenUcUbjD5wbRUDEZoxVaj5YpEVtPgypLoPrhRHQcUsyfZTyzzKo

193 **Goodman S. 2009.** Tests find more than 200 chemicals in newborn umbilical cord blood. *Scientific American* Dec 2
scientificamerican.com/article/newborn-babies-chemicals-exposure-bpa/?redirect=1

194 **EWG. 2009.** Pollution in minority newborns. *Environmental Working Group* Nov 23
ewg.org/research/minority-cord-blood-report/bpa-and-other-cord-blood-pollutants

195 **Daley J. 2018.** The great pacific garbage patch. *Smithsonian* March 23
smithsonianmag.com/smart-news/great-pacific-garbage-patch-larger-and-chunkier-we-thought-180968580

196 **CBD. 2019.** Ocean plastics pollution. *Center for Biological Diversity* May 16
biologicaldiversity.org/campaigns/ocean_plastics

197 **Selin H, Cowling R. 2018.** Cargo ships are the world's biggest polluters. *The Conversation* Dec 19
inverse.com/article/51897-cargo-ships-are-emitting-boatloads-of-carbon-and-nobody-wants-to-take-the-blame

198 **Selin H, Cowling R. 2018.** Cargo ships are the world's biggest polluters. *The Conversation* Dec 19
inverse.com/article/51897-cargo-ships-are-emitting-boatloads-of-carbon-and-nobody-wants-to-take-the-blame

199 **Selin H, Cowling R. 2018.** Cargo ships are the world's biggest polluters. *The Conversation* Dec 19
inverse.com/article/51897-cargo-ships-are-emitting-boatloads-of-carbon-and-nobody-wants-to-take-the-blame

200 **Troeger N, Wieser H, Hübner R. 2017.** Patterns of consumer use and reasons for replacing durable goods. *Researchgate* Feb
doi.org/10.15501/978-3-86336-914-9_5

201 **Simon F. 2019.** Big oil's 'double speak' on climate exposed. *Euractiv* Mar 22

euractiv.com/section/energy/news/big-oils-double-speak-on-climate-exposed-in-new-report

202 **Laville S. 2019.** Top oil firms spending millions lobbying to block climate change policies. *The Guardian* Mar 22 theguardian.com/business/2019/mar/22/top-oil-firms-spending-millions-lobbying-to-block-climate-change-policies-says-report

203 **Simon F. 2019.** Big oil's 'double speak' on climate exposed. *Euractiv* Mar 22 euractiv.com/section/energy/news/big-oils-double-speak-on-climate-exposed-in-new-report

204 **Cook J, Nuccitelli D, Green SA, Richardson M, Winkler B, Painting R, Way R, Jacobs P, Skuce A. 2013.** Quantifying the consensus on anthropogenic global warming in the scientific literature. *Environmental Research Letters* 8:(2) opscience.iop.org/article/10.1088/1748-9326/8/2/024024/meta

205 **Cook J, Oreskes N, Doran PT, Anderegg WRL, Verheggen B, Maibach EW, Carlton JS, Lewandowsky S, Skuce AG, Green SA, Nuccitelli D, Jacobs P, Richardson M, Winkler B, Painting R, Rice K. 2016.** Consensus on consensus: a synthesis of consensus estimates on human-caused global warming. *Environmental Research Letters* 11:(4) doi.org/10.1088/1748-9326/11/4/048002

206 **Milman O. 2017.** Fact check: Trump's paris climate speech claims analyzed. *The Guardian* Jun 2 theguardian.com/environment/ng-interactive/2017/jun/02/presidents-paris-climate-speech-annotated-trumps-claims-analysed

207 **Milman O. 2017.** Fact check: Trump's paris climate speech claims analyzed. *The Guardian* Jun 2 theguardian.com/environment/ng-interactive/2017/jun/02/presidents-paris-climate-speech-annotated-trumps-claims-analysed

208 **Gillis J, Popovich N. 2017.** The U.S. is the biggest carbon polluter in history. *The New York Times* Jun 1 nytimes.com/interactive/2017/06/01/climate/us-biggest-carbon-polluter-in-history-will-it-walk-away-from-the-paris-climate-deal.html

209 **Lavelle M. 2017.** Fossil fuel industries pumped millions into Trump's inauguration. *Inside Climate News* Apr 19 insideclimatenews.org/news/19042017/fossil-fuels-oil-coal-gas-exxon-chevron-bp-donald-trump-inauguration-donations

210 **McCarthy T. Gambino L. 2017.** The Republicans who urged Trump to pull out of Paris deal are big oil darlings. *The Guardian* Jun 1 theguardian.com/us-news/2017/jun/01/republican-senators-paris-climate-deal-energy-donations

211 **McCarthy T. Gambino L. 2017.** The Republicans who urged Trump to pull out of Paris deal are big oil darlings. *The Guardian* Jun 1 theguardian.com/us-news/2017/jun/01/republican-senators-paris-climate-deal-energy-donations

212 **Investopedia. 2019.** How much does it cost to become president? *Investopedia* May 13 investopedia.com/insights/cost-of-becoming-president

213 **Bort R. 2016.** A breakdown of congressional fundraising. *Newsweek* Apr 4 newsweek.com/john-oliver-last-week-tonight-congressional-fundraising-443675

214 **Gilens M, Page BI. 2014.** Testing Theories of American Politics: Elites, Interest Groups, and Average Citizens. *Perspectives on Politics, Cambridge University Press* 12:(3) 564–581 doi.org/10.1017/S1537592714001595

215 **Thunberg G. 2019.** 'Our house is on fire': Greta Thunberg, 16, urges leaders to act on climate, *The Guardian Climate* Jan 25 theguardian.com/environment/2019/jan/25/our-house-is-on-fire-greta-thunberg16-urges-leaders-to-act-on-climate

216 **Lucas C. 2019.** Parliament must declare a climate emergency. *The Guardian* Mar 4 theguardian.com/commentisfree/2019/mar/04/climate-change-emergency-westminster

217 **McKibben B. 2019.** Notes from a remarkable political moment for climate change. *The New Yorker* May 1 newyorker.com/news/daily-comment/notes-from-a-remarkable-political-moment-for-climate-change

218 **UCS. 2019.** Holding fossil fuel companies accountable for nearly 40 years of climate deception and harm. *The Union of Concerned Scientists* May 23 ucsusa.org/global-warming/fossil-fuel-companies-knew-about-global-warming

219 **GP. 2019.** Koch industries: secretly funding the climate denial machine. *Greenpeace* Jun 29 greenpeace.org/usa/global-warming/climate-deniers/koch-industries

220 **BBC. 2019.** Climate change: Ireland declares climate emergency. *BBC* May 9 bbc.com/news/world-europe-48221080?fbclid=IwAR0fdXe4lRPZfH3cjJuyEg8QGN9T7SzQs4UI990ueKYLVBQwRFVlmWDICA4

221 **CED. 2019.** Climate emergency declarations in 623 jurisdictions and local governments cover 83 million citizens. *Climate Emergency Declaration* Jun 16 climateemergencydeclaration.org/climate-emergency-declarations-cover-15-million-citizens

222 **Tong D, Zhang Q, Zheng Y, Caldeira K, Shearer C, Hong C, Qin Y, Davis SJ. 2019.** Committed emissions from existing energy infrastructure jeopardize 1.5°C climate target. *Nature* Jul 1 doi.org/10.1038/s41586-019-1364-3

223 **UN. 2015.** Paris Agreement. *The United Nations* Dec 12 unfccc.int/sites/default/files/english_paris_agreement.pdf

224 **Jackson H. 2019.** National climate emergency declared by house of commons. *Global News* Jun 17 globalnews.ca/news/5401586/canada-national-climate-emergency

225 **BBC. 2019.** Trans Mountain: Canada approves $5.5bn oil pipeline project. *BBC* Jun 18 bbc.com/news/world-us-canada-48641293

226 **Monbiot G. 2019.** Only rebellion will prevent an ecological apocalypse. *The Guardian* Apr 15 theguardian.com/commentisfree/2019/apr/15/rebellion-prevent-ecological-apocalypse-civil-disobedience

227 **UN. 2015.** Paris Agreement. *The United Nations* Dec 12 unfccc.int/sites/default/files/english_paris_agreement.pdf

Printed in Great Britain
by Amazon

54890604R00054